ENERGY SECTOR STANDARD
OF THE PEOPLE'S REPUBLIC OF CHINA

中华人民共和国能源行业标准

Specification for Post-assessment of Occupational Health and Safety for Hydropower Projects

水电工程劳动安全与工业卫生后评价规程

NB/T 10395-2020

Chief Development Department: China Renewable Energy Engineering Institute

Approval Department: National Energy Administration of the People's Republic of China

Implementation Date: February 1, 2021

China Water & Power Press

中国水利水电出版社

Beijing 2024

All rights reserved. No part of this publication may be reproduced, stored in a retrieval system, or transmitted in any form or by any means—electronic, mechanical, photocopying, recording or otherwise, without prior written permission of the publisher.

图书在版编目（CIP）数据

水电工程劳动安全与工业卫生后评价规程：NB/T 10395-2020 = Specification for Post-assessment of Occupational Health and Safety for Hydropower Projects（NB/T 10395-2020）：英文 / 国家能源局发布. -- 北京：中国水利水电出版社，2024.8. -- ISBN 978-7-5226-2724-3

Ⅰ. TV513-65

中国国家版本馆CIP数据核字第2024V5L662号

ENERGY SECTOR STANDARD
OF THE PEOPLE'S REPUBLIC OF CHINA
中华人民共和国能源行业标准

Specification for Post-assessment of Occupational Health
and Safety for Hydropower Projects
水电工程劳动安全与工业卫生后评价规程
NB/T 10395-2020
（英文版）

Issued by National Energy Administration of the People's Republic of China
国家能源局　发布
Translation organized by China Renewable Energy Engineering Institute
水电水利规划设计总院　组织翻译
Published by China Water & Power Press
中国水利水电出版社　出版发行
　　Tel: (+ 86 10) 68545888　68545874
　　sales@mwr.gov.cn
　　Account name: China Water & Power Press
　　Address: No.1, Yuyuantan Nanlu, Haidian District, Beijing 100038, China
　　http://www.waterpub.com.cn
中国水利水电出版社微机排版中心　排版
北京中献拓方科技发展有限公司　印刷
184mm×260mm　16开本　2印张　63千字
2024年8月第1版　2024年8月第1次印刷
Price（定价）：￥335.00

Introduction

This English version is one of China's energy sector standard series in English. Its translation was organized by China Renewable Energy Engineering Institute authorized by National Energy Administration of the People's Republic of China in compliance with relevant procedures and stipulations. This English version was issued by National Energy Administration of the People's Republic of China in Announcement [2023] No. 5, dated October 11, 2023.

This version was translated from the Chinese Standard NB/T 10395-2020, *Specification for Post-assessment of Occupational Health and Safety for Hydropower Projects*, published by China Water & Power Press. The copyright is reserved by National Energy Administration of the People's Republic of China. In the event of any discrepancy in the implementation, the Chinese version shall prevail.

Many thanks go to the staff from the relevant standard development organizations and those who have provided generous assistance in the translation and review process.

For further improvement of the English version, any comments and suggestions are welcome and should be addressed to:

China Renewable Energy Engineering Institute
No. 2 Beixiaojie, Liupukang, Xicheng District, Beijing 100120, China
Website: www.creei.cn

Translating organization:

POWERCHINA Zhongnan Engineering Corporation Limited

Translating staff:

YE Yuxin   LIU Xiaofen   GUO Yulan   LI Qian

Review panel members:

LI Zhongjie	POWERCHINA Northwest Engineering Corporation Limited
QIE Chunsheng	Senior English Translator
ZHANG Ming	Tsinghua University
YE Bin	POWERCHINA Huadong Engineering Corporation Limited
LIU Qing	POWERCHINA Northwest Engineering Corporation Limited

GAO Yan	POWERCHINA Beijing Engineering Corporation Limited
LI Qian	POWERCHINA Chengdu Engineering Corporation Limited
LI Shisheng	China Renewable Energy Engineering Institute

National Energy Administration of the People's Republic of China

翻译出版说明

本译本为国家能源局委托水电水利规划设计总院按照有关程序和规定，统一组织翻译的能源行业标准英文版系列译本之一。2023 年 10 月 11 日，国家能源局以 2023 年第 5 号公告予以公布。

本译本是根据中国水利水电出版社出版的《水电工程劳动安全与工业卫生后评价规程》NB/T 10395—2020 翻译的，著作权归国家能源局所有。在使用过程中，如出现异议，以中文版为准。

本译本在翻译和审核过程中，本标准编制单位及编制组有关成员给予了积极协助。

为不断提高本译本的质量，欢迎使用者提出意见和建议，并反馈给水电水利规划设计总院。

地址：北京市西城区六铺炕北小街 2 号
邮编：100120
网址：www.creei.cn

本译本翻译单位：中国电建集团中南勘测设计研究院有限公司

本译本翻译人员：叶雨欣　刘小芬　郭玉兰　李　倩

本译本审核人员：

李仲杰　中国电建集团西北勘测设计研究院有限公司

郄春生　英语高级翻译

张　明　清华大学

叶　彬　中国电建集团华东勘测设计研究院有限公司

柳　青　中国电建集团西北勘测设计研究院有限公司

高　燕　中国电建集团北京勘测设计研究院有限公司

李　茜　中国电建集团成都勘测设计研究院有限公司

李仕胜　水电水利规划设计总院

国家能源局

Announcement of National Energy Administration of the People's Republic of China [2020] No. 5

National Energy Administration of the People's Republic of China has approved and issued 502 energy sector standards including *Technical Code for Real-Time Ecological Flow Monitoring Systems of Hydropower Projects* (Attachment 1) and the English version of 35 energy sector standards including *Series Parameters for Horizontal Hydraulic Hoist (Cylinder)* (Attachment 2).

Attachments: 1. Directory of Sector Standards
2. Directory of English Version of Sector Standards

National Energy Administration of the People's Republic of China

October 23, 2020

Attachment 1:

Directory of Sector Standards

Serial number	Standard No.	Title	Replaced standard No.	Adopted international standard No.	Approval date	Implementation date
...						
11	NB/T 10395-2020	Specification for Post-assessment of Occupational Health and Safety for Hydropower Projects			2020-10-23	2021-02-01
...						

Foreword

According to the requirements of Document GNZTKJ [2018] No. 100 issued by the General Affairs Department of National Energy Administration of the People's Republic of China, "Notice on Releasing the Development and Revision Plan and the English Translation and Publication Plan of the Energy Sector Standards in 2018", and after extensive investigation and research, summarization of practical experience, and wide solicitation of opinions, the drafting group has prepared this specification.

The main technical contents of this specification include: basic requirements; assessment on conditions of main structures, electromechanical equipment, hydraulic steel structures and slopes; assessment on conditions of safety devices and measures; assessment on control of hazardous factors in workplaces; safety management assessment; assessment conclusions and suggestions; and requirements for preparing a post-assessment report.

National Energy Administration of the People's Republic of China is in charge of the administration of this specification. China Renewable Energy Engineering Institute has proposed this specification and is responsible for its routine management. Energy Sector Standardization Technical Committee on Hydropower Investigation and Design is responsible for the explanation of specific technical contents. Comments and suggestions in the implementation of this specification should be addressed to:

China Renewable Energy Engineering Institute
No. 2 Beixiaojie, Liupukang, Xicheng District, Beijing 100120, China

Chief development organizations:

China Renewable Energy Engineering Institute

POWERCHINA Zhongnan Engineering Corporation Limited

Participating development organizations:

HYDROCHINA Engineering Consulting Co., Ltd.

POWERCHINA Northwest Engineering Corporation Limited

STATE GRID Xin Yuan Group Co., Ltd.

China Southern Power Grid Peak Shaving and Frequency Modulation Power Generation Co., Ltd.

HUANENG Lancang River Hydropower Inc.

China Huadian Corporation Ltd.

HUADIAN Yunnan Power Generation Co., Ltd.

Yalong River Hydropower Development Co., Ltd.

Guizhou Wujiang Hydropower Development Co., Ltd.

Dadu River Hydropower Development Co., Ltd.

SPIC Huanghe Hydropower Development Co., Ltd.

SPIC Wuling Power Corporation Ltd.

China Three Gorges Corporation

Chief drafting staff:

ZHANG Xiaoguang	DU Xiaohu	DAICHEN Mengzi	ZHANG Xiaoli
PAN Jian	ZHU Zhe	YUAN Dongcheng	WANG Jilin
ZHAO Xinchang	JIA Chao	ZHANG Yan	ZENG Hui
MENG Tao	XU Kai	LI Zheng	HU Yong
WEN Shengyuan	LI Yinghuang	DUAN Chuan	ZHANG Xiaoguang
YANG Hong	CAO Yuanyuan	WU Xi	LIU Yunfeng
LI Jing	LI Hua	XU Bin	WANG Yongfei
MA Yimin	LEI Huiguang	ZHAO Dihua	YANG Ping
LIU Aijuan	XIONG Wenqing	YAO Yunlong	

Review panel members:

YANG Zhigang	HE Haiyuan	LIU Shihuang	LIN Peng
XU Jun	FENG Zhenqiu	CHEN Nengping	LIU Changwu
LI Deyu	CHEN Wenbin	MU Kun	CHENG Yunsheng
HU Haitao	SHENG Yongzhen	ZHAO Yi	LIU Rongli
LI Shisheng			

Contents

1 General Provisions ·· 1
2 Terms ·· 2
3 Basic Requirements ·· 3
4 Assessment on Conditions of Main Structures, Electromechanical Equipment, Hydraulic Steel Structures and Slopes ·· 4
5 Assessment on Conditions of Safety Devices and Measures ·· 5
6 Assessment on Control of Hazardous Factors in Workplaces ·· 8
7 Safety Management Assessment ·· 9
8 Assessment Conclusions and Suggestions ································ 10
9 Requirements for Preparing a Post-assessment Report ······ 11
Appendix A Contents of an Occupational Health and Safety Post-assessment Report for Hydropower Project ···· 12
Appendix B Format of the Cover Page of an Occupational Health and Safety Post-assessment Report for Hydropower Project ·· 14
Appendix C Format of the Title Page of an Occupational Health and Safety Post-assessment Report for Hydropower Project ·· 15
Explanation of Wording in This Specification ································ 17

1　General Provisions

1.0.1　This specification is formulated with a view to standardizing the post-assessment of occupational health and safety for hydropower projects.

1.0.2　This specification is applicable to the post-assessment of occupational health and safety for hydropower projects.

1.0.3　The initial post-assessment of occupational health and safety for a hydropower project shall be conducted 2 years after the completion acceptance of occupational health and safety. The time interval of follow-up assessment may be determined and adjusted according to the safety risk of the hydropower project, but should not exceed 5 years.

1.0.4　In addition to this specification, the post-assessment of occupational health and safety for hydropower projects shall comply with other current relevant standards of China.

2 Terms

2.0.1 occupational health and safety

work field aimed to safeguard the health and safety of workers in workplaces and corresponding measures taken in terms of law, technology, equipment, organizational system, education, etc.

Synonyms: occupational safety and health, labor safety and industrial hygiene, labor safety and occupational health

2.0.2 post-assessment of occupational health and safety

activities to make safety assessment of main structures, electromechanical equipment, hydraulic steel structures, slopes, safety devices and measures, hazardous factors in workplaces, safety management, etc. after the hydropower project is put into operation

2.0.3 safety devices and measures

generic term for equipment, facilities, installations, buildings (structures) and other technical measures used to prevent work safety-related accidents

3 Basic Requirements

3.0.1 Post-assessment of occupational health and safety for hydropower projects shall carry out analysis and assessment on main structures, electromechanical equipment, hydraulic steel structures, slopes, safety devices and measures, hazardous factors in workplaces, safety management, etc., put forward safety countermeasures and suggestions, and make assessment conclusions.

3.0.2 Post-assessment of occupational health and safety for hydropower projects shall follow the principles of independence, impartiality, objectivity and scientificity.

3.0.3 The basis for post-assessment of occupational health and safety for a hydropower project shall include the following:

1 Data on completion acceptance of occupational health and safety.

2 Previous reports on post-assessment of occupational health and safety.

3 Other project-related data.

3.0.4 Post-assessment of occupational health and safety for a hydropower project shall be undertaken by an organization with corresponding technical competence.

3.0.5 For the post-assessment of occupational health and safety for a hydropower project, an assessment team shall be established to develop a post-assessment program and conduct on-site investigation, data collection, compliance check, analysis and assessment.

3.0.6 Post-assessment of occupational health and safety for hydropower projects should adopt qualitative and quantitative methods. The safety checklist method, expert on-site inquiry and observation method, comparative analysis method, investigation method, comprehensive indicators system evaluation method, etc. may be used.

3.0.7 For the post-assessment of occupational health and safety for a hydropower project, a post-assessment report shall be provided and shall pass technical review.

4 Assessment on Conditions of Main Structures, Electromechanical Equipment, Hydraulic Steel Structures and Slopes

4.0.1 For the post-assessment of occupational health and safety for a hydropower project, the performance of main structures, electromechanical equipment and hydraulic steel structures shall be assessed based on the project monitoring data, patrol inspection reports, periodic inspection reports of the dam, and operation reports of the power plant, as well as on-site inspection results.

4.0.2 Assessment on the conditions of main structures should cover the following:

1 Water retaining structures.

2 Water release structures.

3 Water conveyance structures.

4 Powerhouses.

5 Navigation structures.

6 Fish passage facilities.

4.0.3 Assessment on the performance of electromechanical equipment and hydraulic steel structures shall cover the following:

1 Turbine-generator units and their ancillaries; power transformation and distribution system; earthing system; lightning protection system; lighting system; control, protection and communication system; oil, compressed air and water supply equipment and systems of the whole power plant; ventilation and air conditioning system.

2 Hydraulic steel structures such as gates, valves, trash racks, hoisting equipment, and penstocks.

4.0.4 Assessment on the slope conditions shall cover the slopes in the hydropower complex area and the reservoir bank slopes near the dam.

5 Assessment on Conditions of Safety Devices and Measures

5.0.1 Assessment on the conditions of flood control and anti-flooding facilities shall cover the following:

1 Hydrologic telemetry and forecasting system.

2 Anti-flooding facilities for all kinds of passages leading to the outside of the buildings.

3 Reservoir emergency drawdown capacity and facilities.

4 Measures for emergency shutdown of turbine-generator units, measures for closing of emergency gates, butterfly valves, spherical valves or ring gates, and water level signaling devices and alarm system used for preventing the powerhouse from being flooded.

5 Drainage system.

6 Power supply system for flood control and anti-flooding facilities.

7 Safety signs and alarm and broadcasting system for downstream river course.

8 Video surveillance system for key flood control and anti-flooding areas.

5.0.2 Assessment on the conditions of fire protection facilities shall cover the following:

1 Fire partitions and emergency evacuation routes in the power plant.

2 Fire water supply and fire extinguishing facilities.

3 Smoke exhaust system.

4 Firefighting power distribution and emergency power supply, emergency lighting, evacuation signs, and automatic fire alarm system.

5 Fire safety signs.

5.0.3 Assessment on the conditions of electrical injury prevention devices and measures shall cover the following:

1 Electrical safety clearance and electric shock prevention measures for high-voltage switchgear.

2 Electric shock prevention measures for low-voltage switchgear.

3 Lightning protection and anti-static earthing system.

4 Electrical explosion-proof measures.

5 Mobile electric tools and insulation tools.

6 PE and PEN signs and safety warning boards.

5.0.4 Assessment on the conditions of mechanical injury prevention devices and measures shall cover the following:

1 Safety clearance in workplaces.

2 Mechanical equipment protective devices.

3 Mechanical equipment protection devices.

4 Warning devices.

5.0.5 Assessment on fall protection devices and measures shall cover the following:

1 Safety devices and measures such as fixed guardrails set on the dam crest and the free face of spillway.

2 Fall protection devices and measures for gate shafts, sumps, shafts, hatches and other parts of hydraulic structures.

3 Fall protection devices and measures for operation or maintenance platforms and passages.

4 Fixed vertical or inclined steel ladders.

5 Fall protection devices and measures for walkways along powerhouse bridge crane rails.

6 Fall protection devices and measures for observation piers, observation accesses, and accesses to high and steep slopes.

5.0.6 Assessment on traffic safety devices and measures shall cover the following:

1 Setting of traffic safety signs and markings.

2 Setting of pavement markings and sight guidance signs.

3 Road lighting facilities and free face protection facilities in the powerhouse and dam areas.

4 Firefighting, lighting and ventilating facilities in tunnels.

5 Measures against road skid, landslide or rockfall.

6 Operation mode and safeguarding measures for navigation structures.

5.0.7 Assessment on safety devices and measures for special-purpose equipment shall cover the following:

1 Inspection and testing of special-purpose equipment and the equipment required for mandatory inspection.

2 Safety conditions of special-purpose equipment.

5.0.8 Assessment on the conditions of safety devices and measures shall also cover natural disaster precautions, emergency facilities, and public security and counter-terrorist facilities.

6 Assessment on Control of Hazardous Factors in Workplaces

6.0.1 Assessment on the control of hazardous factors in workplaces of a hydropower project shall be carried out on the basis of identification, analysis and detection of hazardous factors in workplaces.

6.0.2 Detection of hazardous factors in workplaces shall cover the main workplaces in the project area. The detection parameters should include noise, vibration, air temperature, relative humidity, illuminance, dust, radon concentration in the air, power frequency field intensity, ultraviolet intensity, oxygen content in the air, etc.

6.0.3 Assessment on the control of hazardous factors in workplaces shall cover the following:

1   Noise and vibration.

2   Ventilation, temperature, and humidity.

3   Daylighting and lighting.

4   Dust and toxic and hazardous substances.

5   Electromagnetic radiation.

6   Radon and other radioactive substances.

7   Ultraviolet radiation and high altitude hypoxia.

8   Other hazardous factors.

7 Safety Management Assessment

7.0.1 Safety management assessment of a hydropower project shall be carried out in accordance with current laws and regulations and sector management regulations, taking into account the current situation of safety management of the hydropower project operation organization.

7.0.2 Safety management assessment of a hydropower project shall cover the following:

1 Formulation and fulfillment of work safety objectives.

2 Setup and staffing of a work safety management organization.

3 Establishment and implementation of work safety responsibility system.

4 Establishment and implementation of work safety rules and regulations.

5 Input and use of work safety costs.

6 Implementation of work safety education and training.

7 Management of safety devices and measures.

8 Risk classification and control, and hazard identification and control.

9 Occupational health and occupational disease prevention and control.

10 Emergency management.

11 Investigation and handling of accidents.

12 Safety culture development.

7.0.3 Safety management assessment of a hydropower project shall also cover public security and counter-terrorist management.

8 Assessment Conclusions and Suggestions

8.0.1 For the post-assessment of occupational health and safety for a hydropower project, assessment conclusions shall be made based on the assessment results, and countermeasures and suggestions shall be put forward.

8.0.2 The conclusions of the post-assessment of occupational health and safety for a hydropower project shall cover the following:

1. Conditions of main structures, electromechanical equipment, hydraulic steel structures and slopes.
2. Conditions of safety devices and measures.
3. Control of hazardous factors in workplaces.
4. Safety management, emergency management, and public security and counter-terrorist management.

9 Requirements for Preparing a Post-assessment Report

9.0.1 The occupational health and safety post-assessment report for a hydropower project shall comprehensively and generally reflect all the work of the post-assessment process, and shall have concise and precise text, reliable data and clear topics to facilitate reading and reviewing.

9.0.2 The format and media of the occupational health and safety post-assessment report for a hydropower project shall meet the following requirements:

1. The report shall consist of a cover page, title page, foreword, contents, main body, attachments and attached drawings. The contents of an occupational health and safety post-assessment report for hydropower project should be in accordance with Appendix A of this specification. The cover page of an occupational health and safety post-assessment report for hydropower project should be in accordance with Appendix B of this specification. The title page of an occupational health and safety post-assessment report for hydropower project should be in accordance with Appendix C of this specification.

2. The post-assessment report shall be available in both paper and electronic forms.

9.0.3 The attachments and attached drawings of the occupational health and safety post-assessment report for a hydropower project shall meet the following requirements:

1. The attachments should include a special-purpose equipment ledger, employee certificate ledger, directory of safety management system documents, directory of emergency plan system documents, post-assessment on-site inspection comments and rectification feedback form, etc. The attachments shall be authentic, accurate, complete and valid.

2. The attached drawings should include the project geographical location diagram, general layout, complex management zoning map, plans and sections of main buildings (structures), main electrical connection map, surrounding emergency resources map, and emergency evacuation map. The attached drawings shall be signed as required.

Appendix A Contents of an Occupational Health and Safety Post-assessment Report for Hydropower Project

1   General

1.1   Assessment purpose, scope, work process and procedure

1.2   Assessment basis

2   Construction and operation of the project

2.1   Project profile

2.2   Project operation

2.3   Operation of main safety devices and measures

3   Assessment on conditions of main structures, electromechanical equipment, hydraulic steel structures and slopes

3.1   Water retaining structures

3.2   Water release structures

3.3   Water conveyance structures

3.4   Powerhouses

3.5   Navigation structures

3.6   Fish passage facilities

3.7   Main electromechanical equipment

3.8   Hydraulic steel structures

3.9   Slopes

3.10 Assessment summary

4   Assessment on conditions of safety devices and measures

4.1   Flood control and anti-flooding facilities

4.2   Fire protection facilities

4.3   Electrical injury prevention devices and measures

4.4   Mechanical injury prevention devices and measures

4.5   Fall protection devices and measures

4.6   Traffic safety devices and measures

4.7	Safety devices and measures for special-purpose equipment	
4.8	Natural disaster prevention measures	
4.9	Emergency facilities	
4.10	Public security and counter-terrorist facilities	
4.11	Assessment summary	
5	**Assessment on control of hazardous factors in workplaces**	
5.1	Control of noise and vibration	
5.2	Control of ventilation, temperature, and humidity	
5.3	Daylighting and lighting	
5.4	Dust control and protection against toxic and hazardous substances	
5.5	Protection against electromagnetic radiation	
5.6	Protection against radon and other radioactive hazards	
5.7	Protection against ultraviolet radiation and high altitude hypoxia	
5.8	Control of other hazardous factors	
5.9	Assessment summary	
6	**Safety management assessment**	
6.1	Work safety management organization and staffing	
6.2	Work safety management system	
6.3	Routine safety management	
6.4	Assessment summary	
7	**Conclusions and suggestions**	
8	**Attachments and attached drawings**	
8.1	Attachments	
8.2	Attached drawings	

Appendix B Format of the Cover Page of an Occupational Health and Safety Post-assessment Report for Hydropower Project

B.0.1 The cover page shall contain the following:

1 Name of client.

2 Name of project.

3 Title.

4 Name of assessment organization.

5 Date of completion.

B.0.2 The title shall be uniformly written as " Occupational Health and Safety Post-assessment Report".

B.0.3 The cover page should be in accordance with Figure B.0.3.

Name of Client (SimSun, 22 point, bold)

Name of Project (SimSun, 22 point, bold)

Occupational Health and Safety Post-assessment Report

(SimHei, 26 point, bold)

Name of Assessment Organization

(SimSun, 22 point, bold)

Date of Completion (SimSun, 16 point, bold)

Figure B.0.3 Template of cover page

Appendix C Format of the Title Page of an Occupational Health and Safety Post-assessment Report for Hydropower Project

C.0.1 The title page should consist of two pages. The first page contains the names of main responsible persons of the assessment organization, such as legal representative, technical director, and project director, and the date of completion and the place for the official seal of the assessment organization at the bottom. The second page indicates the assessment personnel with personal signature.

C.0.2 The title page should be in accordance with Figure C.0.2-1 and Figure C.0.2-2.

Name of Client (SimSun, 16 point, bold)

Name of Project (SimSun, 16 point, bold)

Occupational Health and Safety Post-assessment Report

(SimSun, 22 point, bold)

Legal Representative: (SimSun, 14 point)

Technical Director: (SimSun, 14 point)

Project Director: (SimSun, 14 point)

Date of Completion (SimSun, 12 point, bold)

(Official Seal of Assessment Organization)

Figure C.0.2-1 Template of first title page

Assessment Personnel (SimSun, 16 point, bold)

Assessment personnel	Name	Specialty	Technical title	Signature
Project director				
Drafter				
Checker				
Reviewer				

NOTE This table shall be subject to adjustment according to the actual number of participants in the project.

Figure C.0.2-2   Template of second title page

Explanation of Wording in This Specification

1 Words used for different degrees of strictness are explained as follows in order to mark the differences in executing the requirements in this specification.

    1) Words denoting a very strict or mandatory requirement:

    "Must" is used for affirmation; "must not" for negation.

    2) Words denoting a strict requirement under normal conditions:

    "Shall" is used for affirmation; "shall not" for negation.

    3) Words denoting a permission of a slight choice or an indication of the most suitable choice when conditions permit:

    "Should" is used for affirmation; "should not" for negation.

    4) "May" is used to express the option available, sometimes with the conditional permit.

2 "Shall meet the requirements of…" or "shall comply with…" is used in this specification to indicate that it is necessary to comply with the requirements stipulated in other relative standards and codes.